BRIDGES

STEVEN A. OSTROW

MetroBooks

An Imprint of Friedman/Fairfax Publishers

© 1997 by Michael Friedman Publishing Group, Inc.

Library of Congress Cataloging-in-Publication Data.

Ostrow, Steven A.
    Bridges/Steven A. Ostrow.
        p.   cm.
    Includes bibliographical references and index.
    ISBN 1-56799-445-8
    1. Bridges.   I. Title.
TG145.078   1997
624'.2--dc21   96-40509

Editor: Tony Burgess
Art Director: Kevin Ullrich
Designer: Galen Smith
Photography Editor: Kathryn Culley
Production Manager: Jeanne Hutter

Color separations by HK Scanner Arts Int'l Ltd.
Printed in the United Kingdom by Butler & Tanner Limited

1   3   5   7   9   10   8   6   4   2

For bulk purchases and special sales, please contact:
Friedman/Fairfax Publishers
Attention: Sales Department
15 West 26th Street
New York, NY 10010
212/685-6610 FAX 212/685-1307

Visit our website:
http://www.metrobooks.com

For my parents, Vicki and Jack, who are

celebrating their fiftieth wedding anniversary

as I write this book.

CONTENTS

INTRODUCTION

STONE ARCH CONSTRUCTION

Technology has always been the driving force of innovation in bridge building and design. The Romans made the first leap forward in building materials, changing from timber to more durable stone. The best structural form that could support stone was the half-circle arch. The original innovators of this arch seem to have been the Etruscans, who lived in the central plains of Italy. These designers used wedgelike, tapered stone blocks, called voussoirs, to construct their arches. After conquering the Etruscans, the Romans absorbed their best art and technology. The half-circle arch bridge was used effectively by the Romans for the next five centuries. Many of these bridges are still standing.

The greatest construction problem confronting the Romans was building a suitable foundation to support the weight of the stone bridge and the traffic going across. The foundation is the most important part of the bridge. Before they could even build the foundation, however, they had to first divert the river. Earlier civilizations built bridges by temporarily diverting the river and working directly on the dry riverbed, or by dropping stones into the water until they massed on the bottom of the riverbed, or by utilizing available rock outcroppings to construct bridge piers.

To solve their problem, the Romans developed cofferdams: timber piles were driven into the riverbed to form a box where a bridge pier was to be located; the water was then drained out to create a dry enclosure suitable for building purposes. This method of building resulted in bridges that have stood for centuries and enabled the Romans to construct foundations in the middle of practically any waterway.

The Roman Empire stretched across Europe and into modern-day Turkey and beyond. Remains of Roman military roads, bridges, and aqueducts can be found throughout these regions to this day. In the Rhineland, for instance, Julius Caesar constructed the one-thousand-foot (304.8m) timber trestle Rhine Bridge; the building of this bridge is discussed in his campaign memoirs, *De Bello Gallico*. Great timber trestles were driven into the riverbed, and logs were laid across the top to make a roadway. Caesar claimed that this impressive structure was completed in just ten days, a record of which any modern army would be proud.

OPPOSITE: *The first manmade bridges were stones or tree trunks placed across streams. One of the earliest examples known is the "clapper" or "clam" bridge over Walla Brook in Dartmoor, England. Outcroppings of flat stones (alternatively called clams or clappers) were used in spans up to fifteen feet. The slabs are similar to the horizontal beams of Stonehenge.* RIGHT: *This ancient Roman bridge at San Martino, Italy, is an excellent example of what the Roman engineers acheived.*

The most important Roman bridges were built of stone, with stone piers, stone roadways, or both. One famous example, constructed of stone piers with a timber span, is Trajan's Bridge, which was built over the Danube in A.D. 106 by Appollodorus of Damascus. The ruins of the bridge's stone piers were still visible in the late 1900s. An illustration of Trajan's Bridge is found on a column relief in his tomb. Although it is out of scale, this relief provides a rich tapestry of detail. The bridge had a latticed parapet roadway, a span made of timber arches, and cut stone piers.

In Dion Cassius' chronicles of the history of Rome, which appeared around A.D. 200, Trajan's Bridge is described as having twenty massive stone piers measuring 150 feet (45.7m) tall, fifty feet (15.2m) thick, and sixty feet (18.2m) wide. The 110 feet (33.5m) between each pair of piers was spanned by timber arches. The column relief in Trajan's tomb shows that the timber spans were further supported by three arches, each with a monumental drawbridge and a fortified citadel erected at each end of the bridge. According to Cassius, this massive and complex structure was built by partially diverting the river and constructing cofferdams to lay dry foundations.

Engraved on a surviving stone pier of Trajan's Bridge is the following inscription: "The Roman Augustus, truly a great bridge-builder, constructed this bridge that even the grand rapids of the mighty Danube cannot overcome." Unfortunately, this inscription

Located near Roquefavour, France, this fine example of the ancient Roman aqueduct stands more than one hundred feet (30.4m) high. Aqueducts were built to carry water to the Romans' ubiquitous baths and fountains. Roman engineers were kept busy providing the growing empire with the necessary infrastructure of roads, bridges, and aqueducts.

did not save Trajan's Bridge from the hands of the Romans themselves. In order to prevent invaders from using the bridge, the Romans destroyed the timber spans less than thirty years after their construction.

In England, traces of the first timber bridge across the Thames were found at Ermine and Watling Streets, and excavations of the Trent River in the late 1880s uncovered massive, diamond-shaped masonry piers running diagonally across that river. Rather than build a bridge at the Trent's narrowest point, which would have required new roads, the Romans built a longer bridge to connect existing roads. This shows the confidence the Romans had in their bridge building—they were willing and able to build a longer structure to reach the farther shore. The excavated piers were so solidly built that they had to be blown up in the nineteenth century in order to make way for a new bridge across the Trent.

The strength of the Roman piers was the result of a closely held technological secret: a natural cement named *pozzolana* found at Pozzuoli, near Naples. When mixed with lime, this volcanic sand creates a hydraulic cement comparable to modern-day concrete. The Romans used this concrete in foundations and piers submerged under water. The Romans were also fine brick makers, and they manufactured large, flat one- to one-and-a-half-inch (2.5 to 3.8cm) clay, earth, and sand tiles. These bricks were dried for long periods of time to create a material that was both durable and structurally sound.

In building their half-circle arches, the Romans built on a grand scale, with spans extending more than 130 feet (39.6m). The half-circle shape meant that each voussoir was exactly the same shape. These arch stones, which required no mortar between them, sometimes weighed more than seven tons (6.35t) each and often needed to be raised more than one hundred feet (30.5m) into the air, then supported and secured with the next stone—a daunting task.

The first bridge across the Tiber River in Rome was the Pons Sublicius, constructed entirely of wood. The name of the bridge is derived from the Latin word *sublicae*, meaning "wooden beams." The exact location of the bridge has never been determined; it was washed away in a flood in A.D. 69. Wood was used in an effort to appease the gods—the Romans believed that the river gods would take offense if a permanent bridge were made of stone. As a further gesture of appeasement, the Romans instituted an annual human sacrifice to the Tiber River. Eventually this ritual evolved into the casting of reed dummies into the Tiber.

Centuries later, when the Romans rebuilt this bridge in stone, the roadway was still constructed of timber, and no iron was permitted to be used anywhere on the bridge. Iron was used for war, and the river gods would not permit any other use. (In fact, the Romans believed that the other gods would

be angered by the use of iron for any other purpose; they felt so strongly about this that iron tools were not even used in repairing temples.) This restriction on the use of iron in bridge building extended even into nineteenth-century England, when the pier foundations of the new London Bridge were constructed of timber piles.

The oldest bridge still standing in Rome is the Pons (now Ponte) Fabricius. Built in 62 B.C. by Lucius Fabricius, this bridge is also referred to as the Quattro Capi, meaning "the four heads." A figure of Jason with four heads once graced the parapet of this bridge. The Quattro Capi connects what was then known as the Island of Aesculapius to the city proper; the island was the home of the Temple of Aesculapius, the object of many pilgrimages for healing and cures. The island is now known as the Isola Tiberina. The other side of the island was linked to the city by a bridge built around 46 B.C. by Lucius

Pons Fabricius, Rome. The oldest bridge still standing in Rome, the Pons Fabricius was completed in 62 B.C. by the Roman architect Lucius Fabricius. This gem over the Tiber River connects the city of Rome to the Island of Aesculapius (now the Isola Tiberina). The bridge consists of two spans, a central arch built over the midspan pier, and two smaller arches that were sadly removed when the present-day embankments were constructed.

ABOVE: *Ponte Sant'Angelo, or the St. Angel Bridge, in Rome was built in* A.D. *134 by the emperor Hadrian. In Roman times, this bridge over the Tiber River was known as the Aelius Bridge. The iron balustrade and ten angels were added by Giovanni Lorenzo Bernini by order of Pope Clement IX in 1668.* RIGHT: *Two of Bernini's angels on the Ponte Sant'Angelo.*

Cestius, a governor of Rome. The Pons Cestius and the Pons Fabricius both formed dramatic gateways to the island, which was fashioned to look like a large ship. The island banks were lined with travertine and an obelisk reminiscent of a ship's mast jutted up in the middle of the island.

The emperor Hadrian, an amateur architect and bridge builder, left an enviable legacy. To control trade and enforce their rule in the regions they had conquered across Europe, the Romans built walls in Great Britain, Germany, and Rumania. Around A.D. 120, they constructed a stone wall that measured ten feet (3m) wide and twenty feet (60.9m) high and extended 73 miles (117.5km) from the Tyne River to the Solway Firth in Great Britain. There is evidence that this wall, called Hadrian's Wall, actually crossed over the Tyne on a Roman bridge made of timber.

Several years later—in A.D. 134, to be precise—Hadrian built a bridge in Rome. The extensive foundation of this bridge, the Aelius Bridge, has survived to this day. The bridge (now named Sant'Angelo) was rebuilt on the original piers for Pope Nicholas V and Pope Clement IX. In the seventeenth century,

the sculptor and architect Giovanni Lorenzo Bernini was commissioned to widen the roadway and add a detailed railing with ornamental angels; these details grace the structure to this day.

Spain is also the site of many impressive bridges built by the Romans. The Alcantara Bridge over the Tagus River was built by Caius Julius Lacer during the reign of Emperor Trajan. This structure is still standing after nineteen centuries and is prominently featured in Spanish travel brochures. The Moors named this bridge Al Kantarah, which translates into "the bridge," a compliment in which no superlative is required. Caius Julius Lacer is buried beside his greatest achievement, which rises 170 feet (51.8m) above the river on granite piers that are thirty feet (9.1m) wide and thirty feet (9.1m) deep.

As was the practice, no mortar was utilized in the construction of this bridge. No mortar was needed. The voussoirs, some weighing more than seven tons (6.35t), were lifted 170 feet (51.8m) and set into the 98-foot (29.8m) spans. This bridge has been compared to the Great Pyramids; it has withstood the ravages of a number of wars

and the passing of many centuries. When the French Grande Armée retreated across the bridge in 1812, they destroyed one of the larger spans in an effort to halt the advance of the Duke of Wellington's army. The strategy did not succeed, however, and the pursuing army was able to cross over on a hastily constructed suspension bridge made of rope. In rebuilding the war-ravaged bridge, the Spanish had to resort to the use of mortar, a practice that Roman engineers had never had to follow.

IRON GIRDER CONSTRUCTION

The second great leap forward in bridge building would not occur until the Industrial Revolution of the eighteenth century. This revolution brought with it many changes to transportation, structures, and material. A major impetus of the Industrial Revolution was the newly invented railroad, which created a need for bridges that could span rivers and carry the heavy loads of steam engines and their freight. The only material that

RIGHT: *The Craigellachie Bridge over the Spey River is the oldest surviving cast-iron bridge that utilizes the modern metal-truss form. Completed in 1814 and located at Elgin in the Scottish Highlands, the 150-foot (45.7m) span is one of Thomas Telford's masterpieces. His use of cast iron in this bridge created a new direction in bridge building and a technological breakthrough in the use of materials.*

could support heavy loads and span great distances was iron. Because of prevailing superstitions, however, iron was shunned in bridge building, and it took a "revolution" to free it from its restricted role to become the leading technological advancement of its age.

The first bridge made of cast iron was a semicircular arch supporting a roadway constructed over the Severn River in Coalbrookdale, England, in 1779. Known as the "Iron Bridge," it was the only span that survived the harsh floods of 1795.

Not long after iron began to be used, its strength and value became clear to British engineer Thomas Telford, who began designing and developing new forms in cast iron for use in bridges. Telford's structures spanned rivers without the need for intermediary piers in midstream. This wide-span construction prevented damage brought on by floods and ice and offered boats clear sailing under the spans.

Telford's Bonar Bridge (1810) over the Dornoch Firth in Scotland incorporated a clear span of 150 feet (45.7m) that would normally have required two intermediary midstream piers to support the structure. His breakthrough wide-span design, fabricated of cast iron, was technically superior to the structures built by his contemporaries. Telford's oldest surviving Bonar-style cast-iron arch bridge is the Craigellachie Bridge over the Spey River in Elgin, Scotland. Built in 1814, this bridge is an excellent example of the modern metal-trussed bridge; its "diagonal member" design has been replicated in bridges ever since.

In 1800 Telford submitted a bridge design to replace the crumbling and ancient London Bridge. Had it been built, Telford's six-hundred-foot (182.8m) cast-iron span would have captured the world's attention eighty years ahead of Alexandre-Gustave Eiffel's tower or John A. Roebling's Brooklyn Bridge. This structure would have dominated the London landscape and sent a clear message regarding Britain's technological power and greatness around the world. Instead, Parliament approved an older stone masonry "Parisian-type" bridge design to replace the aging relic. Stone was still seen as the material of choice in bridge construction as late as 1831, when this new London Bridge was opened. Iron bridges were relegated to rural and distant locations where stronger structures and longer spans were required for the burgeoning railroads.

In 1826 Telford completed one of his masterpieces, the bridge over the Menai Straits in Wales. The Menai Bridge's span of 580 feet (176.7m) was the world's longest suspension of its day, and the forerunner of modern bridge design. The structure was suspended by wrought-iron chains and the clear span over the Menai towered above the straits. The success of this bridge led to the replacement of the stone arch by the iron cable in bridge design. Iron was now recognized as a material of strength and durability. Thomas Telford's innovative bridge concepts established him as the father of the iron bridge. His concern for aesthetics, his use of new materials, and his structural daring set him apart from other structural designers of his day.

The mid-nineteenth century saw a frenzy of bridge building necessitated by the surging growth of railroads. Two prominent figures who rose to the top of their profession during this period were Robert Stephenson (son of

George Stephenson, inventor of the world's first viable steam locomotive) and Isambard Kingdom Brunel (son of the famous British engineer Marc Brunel). These men dominated their field and brought British technological might to new heights.

Stephenson's greatest bridge was erected in 1850 over the Menai Straits in Wales. His Britannia Bridge consisted of solid wrought-iron plates put together to form a long hollow rectangular tube. This solid iron tube had trains running through it, similar to a tunnel. Although the design originally included cables, it turned out that the intermediary stone towers provided enough stability, and supplementary suspension cables were not required.

Brunel was a daring, inventive structural engineer who built railroads, railroad stations, and bridges. His finest creation,

besides the Great Western Railway between Bristol and London, was the Royal Albert Bridge at Saltash over the Tamar River. Also called the Saltash Bridge, it had twin spans of 455 feet (138.6m) each. The innovative design combined a tubular arch (oval in cross section) with a wrought-iron cable suspended below. This design was considerably more economical than Stephenson's Britannia Bridge (forty-seven hundred pounds [2,131.5kg] of iron per foot versus seven thousand pounds [3,174.5kg] of iron per foot). Brunel's pure structural form of arch and cable is clearly defined and integrated with a horizontal deck extending below the entire structure.

The age of the iron bridge culminated in the latter part of the 1880s with the creative geniuses of Gustave Eiffel and John A. Roebling. Eiffel's fame has endured and

grown with the stature and elegance of the Parisian tower named for him. His iron-arch tower construction is the same structural design he had developed for his bridges and viaducts. The Rouzat Viaduct, Eiffel's 1869 railroad trellis bridge over the Sioule River, stood on two iron towers. Each tower was over two hundred feet (60.9m) in height,

with the base spreading out to prevent the wind loads from toppling the viaducts. Eiffel's 525-foot (160m) span over the Douro River in Oporto, Portugal, was opened to the railroad traffic of the Royal Portuguese Railway in 1877. Known as the Pia Maria Bridge, this wrought-iron crescent-shaped arch, which rises more than 250 feet (76.2m) into the air, is still standing today. The construction of this bridge confirmed Gustave Eiffel as the leading bridge designer on the European continent.

Following the Pia Maria's successful completion, Eiffel was asked to build a railway viaduct over the Truyère River in St. Flour, France. When Eiffel's wrought-iron Garabit Viaduct was completed in 1884, it was the longest arch bridge in the world. Its graceful crescent form with a continuous wrought-iron trellis girder is Eiffel's bridge masterpiece. In fact, Garabit was the design forerunner of an even greater structural art form: the Eiffel Tower, completed in 1889.

The most prominent nineteenth-century American structural engineer was John Augustus Roebling, the father of the modern-day cable suspension bridge. Roebling was a man of vision who established a wire rope manufacturing plant to fabricate the wire cable he would utilize for his suspension bridge designs.

Roebling's first suspension bridge was a canal bridge built over the Allegheny River in 1844, followed by a roadway span in Pittsburgh, Pennsylvania, over the Monongahela River. Roebling's first widely acclaimed structure was his combination road-and-rail suspension bridge over Niagara Falls. Completed in 1855, this 821-foot (250.2m) iron-wire cable suspension bridge was the only suspension span to successfully carry a railway over an extended period of time. Because the suspension bridge design was not considered favorable for heavy railroad loads, Roebling stiffened the superstructure with wire cable stays radiating from the stone towers, thus effectively preventing dangerous oscillations. This bridge was replaced in 1897, when railroad weight requirements increased substantially and it was no longer economically feasible to maintain the structure.

The prototype of Roebling's greatest masterpiece—the Brooklyn Bridge—was the Cincinnati Bridge over the Ohio River. This 1,057-foot (322.1m) iron-wire cable suspension was the world's longest when it was completed in 1866. Its construction had been delayed for several years but commenced again during the Civil War. This delay seems actually to have been fortuitous in the eventual construction of the Brooklyn Bridge. Initially, in the late 1850s, the citizenry of New York City had rejected Roebling's urgings to span the East River with a suspension bridge. However, the bitterly cold winter of 1866 changed the minds of the residents of both cities (Brooklyn at this time was an independent city) bordering the East River. The ice-choked river prevented ferry commuters from getting to work, and the campaign to build a bridge across the river finally gained civic support. By the time the Brooklyn Bridge was opened to the public on May 24, 1883, the bridge had gained a symbolic national and international importance, and had single-handedly ushered in a new technological era.

Classical
Bridges of
Europe

Early Renaissance bridges in Europe were as popular and as busy as present-day shopping malls. Renaissance stone bridges had rows of shops, homes with cellars, chapels, fortifications, parapets, and drawbridges. During the reign of Queen Elizabeth I, for example, an exclusive residential neighborhood sprang up on the famous London Bridge. The Nonesuch House, built on the bridge, was the residence of young nobles of the queen's court.

For almost six centuries, the London Bridge was the only bridge across the Thames. Work on this bridge began during the reign of King John in 1176. Peter of Colechurch, the bridge master who started this daunting task, died in 1205, four years before the bridge was completed; he was buried in the bridge chapel that was constructed on the span.

The London Bridge had nineteen arches and a drawbridge for defense and ship access. The spans varied from fifteen to thirty-four feet (4.5–10.3m) and were eighteen to twenty-six feet (5.4–7.9m) wide. It was the first bridge with masonry foundations to be built on a swiftly flowing river estuary, and hundreds of workers perished during the thirty-three years of construction.

Old London Bridge was famous for its streets filled with shops, whose rents helped pay for the bridge's upkeep. To raise more funds, tolls were charged for crossing over the narrow, congested artery, where fires and accidents occurred frequently. In 1212, only three years after the bridge opened, a fire destroyed every shop and house on the bridge. After each such disaster, the reconstruction was always bigger and better.

The end for this wondrous structure came at the beginning of the Industrial Revolution, by which time the fires and the ravages of the river had finally taken their toll. In 1831 Old London Bridge was replaced by a new structure designed by John Rennie.

On the Continent, just a year after the London Bridge was begun in 1176, construction began on the first stone bridge to be built in France since the fall of the Roman Empire. Built over the Rhone River at Avignon, this bridge was designed by Bénézet (about whom more will be said in a later chapter), who felt he had a holy calling to accomplish the task. Inspiring the citizens, he raised the necessary funds to construct a series of slender masonry arches across the Rhone, each with a span of one hundred feet (30.4m). Near the Avignon side a chapel was built honoring St. Nicholas, the patron saint of sailors, travelers, bakers, merchants, and children. Bénézet's fame and stature as a bridge architect became so great that after his death he was made the patron saint of bridge builders.

During the Middle Ages, bridges that led to towns (rather than simply connecting two parts of a road across a river) not only made access easier for travelers but also provided each town with a means of defense. Bridges often served as forts protecting the town from raiders. The Pont Valentré at Cahors, France, is an excellent surviving example of this type of medieval bridge. Twin fortified towers rising more than one hundred feet (30.4m) above the river protect the sixteen-foot (4.8m) wide roadway. Because of the sheer size of this structure, it was nearly impossible for raiders to capture the bridge and gain access to the town.

Another famous war bridge was the Vieux Pont, situated in the Pyrenées at Orthez, France. The Vieux Pont entered the annals of military history when, during the Wars of Religion (1562–98), Catholic priests were thrown from the opening in the bridge wall by the invading Huguenots. Consequently, the opening became known as the "Priests' Window." Several centuries later, in 1814, when the French were in full retreat from the Duke of Wellington's forces during the Peninsular Wars, the British were stopped at the Vieux Pont—taking advantage of the bridge tower, forty-five French soldiers held off the pursuing English forces.

The first masonry bridge built in Paris was the Pont Notre Dame, constructed in 1507. Like the timber structure it replaced, the new bridge had two streets filled with homes that occupied the roadway. This allocation of space continued until 1786, when the roadway was widened and the houses eliminated. The old arches of this bridge were entirely demolished in 1853.

In Italy, Florence's famous Ponte Vecchio is a fine example of medieval bridge construction and function. Built in 1345, this historic landmark is little changed from its use and appearance during the reign of the Medicis—it still has rows of goldsmiths and jewelry shops lining its roadway.

In Venice, the most challenging aspect of bridge building was creating a stable foundation. Antonio da Ponte solved this problem in the sixteenth century, when he had more than six thousand timber piles driven sixteen feet (4.8m) into the soft alluvial subsoil at each end of his Rialto Bridge. Da Ponte's bridge is still Venice's busiest canal crossing and it continues to serve the city with grace and elegance as it has since its construction.

The first stone edifice built solely for carriage and pedestrian traffic was the Pont Royal in Paris. Vehicular traffic had increased substantially by the turn of the seventeenth century, and obstructions caused by houses and shops on the roadway could no longer be ignored. As cities expanded throughout Europe, the modern masonry bridge—devoid of all houses and dedicated solely to carrying traffic on wide roadways—gradually evolved.

PAGES 18–19: *In this painting by Kaspar van Wittel, the Ponte Sant'Angelo, leading to the entrance of the Castel Sant'Angelo on the right, is the central focus of interest. Saint Peter's Basilica rises over the bridge in the distance.*

ABOVE: *The Bridge of Sighs is a connecting structure between the Venetian Palace of the Doge (governor) and the prison across the canal. Antonio da Ponte, the builder of the Ponte Rialto, rebuilt the prison and constructed the Bridge of Sighs in 1589. Prisoners would be taken over to the Palace of the Doge for trial and returned via the bridge. With a romantic name that captures the very nature of Venice itself, the Bridge of Sighs, like the Rialto Bridge, is one of the finest sixteenth-century masterpieces in the world.*

The most popular and well-known bridge in Venice, the Rialto Bridge spans the Grand Canal. Following an extensive design competition, Antonio da Ponte was chosen over a number of other eminent candidates, including Michelangelo and Palladio, to build an eighty-eight foot (26.8m) span in the soft alluvial mud that forms the many islands of Venice. Each abutment is supported by more than six thousand six-inch (15.2cm)-diameter timber piles, each driven sixteen feet (4.8m) into the ground. Da Ponte's innovation was to "lay up" a series of wedge-shaped stones over the piles. This innovation was widely adopted and continues to be used to this day.

The Ponte Vecchio, built over the Arno River in 1367, is one of the most enduring and picturesque symbols of Florence. Medieval bridges were typically built with residences, shops, and chapels, and this bridge is lined with shops and stores to this very day. Only a very few fine examples have survived with so few alterations since the era of the Medicis.

The Tall Bridge at Serchio, in Tuscany, is a typical example of the Roman engineering used throughout Europe during the five hundred years of Roman rule. Built on a grand scale, some of the half-circle arches reached a span of 140 feet (42.6m), with columns rising to a height of 160 feet (48.7m). The stones of the great arches were set without mortar or cement, a feat that cannot be matched today.

The Pont Alexandre III—one of the most beautifully ornamental bridges in Paris—was built in 1900 over the Seine River. The three-hinged arch structure was designed by Résal. The towers guarding the ends of the bridge are each topped by a golden Pegasus. The bridge spans more than three hundred feet (91.4m) and honors Emperor Alexander III of Russia (1845–1894), whose reign was distinguished by a close alliance with France.

Built between 1578 and 1604 on the orders of King Henry III, the Pont Neuf was the second Renaissance bridge constructed in Paris. For more than two hundred years, merchants paid rent to maintain their rows of shops and stalls on the bridge, making it both a favorite marketplace and a main artery of the city. During the French Revolution, the Pont Neuf was the famous thoroughfare used to carry victims off to the guillotine.

ABOVE: *Popularly known as the Bridge to Nowhere and as the Pont St. Bénézet, the Pont d'Avignon, built by Bénézet between 1177 and 1187, was one of the first bridges constructed in Europe after the fall of the Roman Empire. Bénézet died in 1184 and was buried in the chapel constructed on the bridge. Shortly after his death Bénézet was made the patron saint of bridge builders.* **OPPOSITE:** *The Pont Charles Albert (de la Caille) spans the Gorge de Caille, a valley in the Haute-Savoie near the border between Switzerland and France. Today the bridge is out of use, but the old castle towers are popular tourist attractions and have been preserved.*

The Pont Valentré was built between 1308 and 1355 over the Lot River in Cahors, France. Crowned by three fortified towers standing guard over the city, the bridge is the finest example of a medieval war bridge that survives in France today. The towers rise to a height of 116 feet (35.5m) and serve as sentinels for the citizens of Cahors.

LEFT: *Spanning the Vltava River in Prague, the Charles Bridge was commissioned by Emperor Charles V in 1357. It was finally completed in 1503, and it still holds the world record for duration of bridge construction. The bridge entry parapet is covered with ornamental figures of saints, and the far end is protected by a medieval Bohemian guard tower.* BELOW LEFT: *The Chain Bridge, built over the Danube River in Budapest at the start of the nineteenth century, is an early example of an eyebar chain suspension bridge. The original structure was destroyed during World War II, but the Hungarian people reconstructed this national symbol in 1949.* OPPOSITE: *The Tower Bridge is a bascule-type drawbridge built over London's Thames River; it opened to river traffic in June 1894. This bridge is short, but impressive—two imposing turreted towers carry a span of only two hundred feet (61m).*

LEFT: *Isambard Kingdom Brunel's first suspension bridge, located at Clifton, England, spans the steep limestone cliffs at the Avon Gorge. Work on the bridge began in 1831 but was soon halted because of riots in nearby Bristol and a subsequent lack of funds. The towers were erected in 1843, but the bridge was not completed until 1864. Brunel, however, never saw the finished bridge—he died in 1859.*

ABOVE: *The Magere Brug, or Meagle Bridge, is a picturesque little structure that crosses the Amstel River in Amsterdam. In a bascule-style drawbridge such as the Magere Brug, the roadway lifts up on attached hinges on both ends in order to permit fishing boats at pass through to the sea.*

OPPOSITE: *The San Martin Bridge, which was heavily influenced by the Moorish style, was built in 1212 in Toledo, Spain. Five arches span the Tagus River at a height of nearly one hundred feet (30.4m) above the river. The St. John of Kings Monastery rises up behind the formidable tower gate of the bridge.* **ABOVE:** *This ancient Roman bridge in the Scottish highlands was made famous by the poet Robert Burns and by the Broadway musical* Brigadoon. *Using gray granite, Roman bridge builders produced a structure that they could replicate virtually anywhere.*

ABOVE: *The Bohemian city of Prague, also known as the City of One Hundred Spires, straddles the Vltava River in the Czech Republic. Many bridges span the Vltava in Prague, of which the Charles Bridge (foreground) is only the most famous. Looking upstream, the three other bridges in this photograph are: Most Legii, or Legions' Bridge, commemorating Czech troops who fought against the Bolsheviks in World War I; Jiraskuv Most, or Jirasek Bridge, named for a renowned Czech writer; and Palackeho Most, or Palacky Bridge, also named for a writer.* **RIGHT:** *The Karl Theodor Bridge in Heidelberg, Germany, is the oldest of three bridges connecting the two sections of the city separated by the Neckar River. The Karl Theodor Bridge was constructed in 1788.*

ROMANCING

THE BRIDGE

BRIDGES ARE MANY THINGS TO MANY PEO-

PLE. BEYOND THE UTILITARIAN FUNCTION

OF PROVIDING PASSAGE FOR VEHICULAR,

RAILROAD, AND PEDESTRIAN TRAFFIC,

BRIDGES SERVE MANY OTHER NEEDS. TO

SOME, BRIDGES MAY SERVE AS NOSTALGIC

REMINDERS OF PAST JOURNEYS OR AS LOCA-

TIONS FOR ROMANTIC RENDEZVOUS. TO

OTHERS, A BRIDGE MAY BE VIEWED AS A

STRUCTURAL WORK OF ART. BRIDGES TOUCH

US EVERY DAY, BUT THEY ARE USUALLY

TAKEN FOR GRANTED AND RARELY GIVEN A

SECOND THOUGHT.

Bridges have always made fine settings for movies. For example, Jimmy Stewart contemplates suicide from a bridge guardrail in the 1946 movie *It's a Wonderful Life*, but fortunately decides against it when he sees what would have occurred had he never lived. In the 1948 gangster movie *Force of Evil*, John Garfield frantically escapes his pursuers as the George Washington Bridge prominently fills the movie screen over his right shoulder. In *My Favorite Year* (1982), Peter O'Toole steals a horse in New York City's Central Park and gallops across the famous Bow Bridge over the park's lake.

The scenic Bow Bridge is the oldest surviving wrought-iron girder bridge in the United States. For support and stiffening, this bridge, designed by Calvert Vaux and Jacob Wrey Mould, has distinctive wrought-iron truss rods, similar to railroad freight car truss rods, that are exposed on the underside of the pedestrian walkway.

Central Park has five other cast-iron pedestrian bridges designed by Vaux and Mould. These bridges constitute the largest and oldest grouping of cast-iron arch bridges

in the United States. The bridges' graceful, decorative art nouveau styling sets a romantic tone for any couple strolling in the park.

Bridges have always had sentimental overtones. In Clint Eastwood's 1995 movie, *The Bridges of Madison County*, the covered bridge served as the perfect bucolic setting for a romantic interlude. The film was shot on location in Iowa; however, the bridge described in the original book is the Roseman Bridge, while the bridge used in the movie is the Holliwell Bridge, selected because the cinematographers deemed it more photogenic.

Today, covered bridges are nostalgic reminders of the sentimental, small-town, or rural past of a nearly lost nineteenth-century North America. However, it was not long ago that covered bridges served primarily utilitarian functions. For example, the longest single-span covered bridge in the world is located in Grass Valley, California. The bridge has a Howe truss-framed arch carrying its 233-foot (71m) span over the South Fork Yuba River. Constructed by local builder David Wood in 1862 for the Virginia City Turnpike Company, the bridge was on the main route serving the Comstock Lode, the silver-mining area of western Nevada.

The longest two-span covered bridge in the world connects the towns of Cornish, New Hampshire, and Windsor, Vermont. Built in 1866 over the Connecticut River by local builders—James F. Tasker of Cornish and Bela J. Fletcher of Claremont, Vermont—the structure cost $9,000 and utilized the lattice truss system patented by Ithial Town. This landmark bridge was originally part of a private toll road and continued as such until purchased by New Hampshire and made toll-free in 1943.

Another unusually designed covered bridge worth mentioning is the Humpback Covered Bridge over Dunlaps Creek, near Covington, Virginia. Looking like a beached whale with an eight-foot (2.4m) sloping crown, this bridge, built just prior to the

Civil War, was reportedly spared destruction during the war when it was expedient for the North and the South to agree that both could continue using it.

The oldest surviving stone bridge in the United States is a double-arched 80½-foot (24.5m) bridge over the Ipswich River in Massachusetts. Built in 1764, it was widened in the nineteenth century to accommodate two lanes of traffic.

Before the coming of the railroad, the number one means of commercial transportation was by canal. Canal boats had to cross rivers in the same way as railroads or roadways, and they did so through the use of bridge aqueducts. In the construction of a bridge aqueduct for a canal, the weight of the water had to be incorporated into the calculations. John Roebling's oldest suspension bridge in the United States is an aqueduct bridge spanning the Delaware river between Lackawaxen, Pennsylvania, and Minisink Ford, New York. Built in 1849 with three midriver piers sitting in the Delaware, this bridge is the technological forerunner of the "Great Bridge" that was to rise over the East River a scant thirty-four years later.

In 1968 Robert McCullock brought the ten-thousand-ton (9,080t) London Bridge to his new Arizona development as a tourist attraction. The idea worked; the bridge now attracts well over one million visitors each year to Havasu City, Arizona. London, meanwhile, is now served by a modern, three-span bridge of prestressed concrete.

PAGES 40–41: The Humpback Covered Bridge is one of the most famous and unique covered bridges in the United States. The bridge was built over Dunlaps Creek near Covington, Virginia, in 1857 by the James River and Kanawha Turnpike; the designer remains unknown. The distinctive humpback curve is formed by an eight-foot (2.4m) rise at the top of the chord.

OPPOSITE BOTTOM: The Old Comstock Covered Bridge in East Hampton, Connecticut, is one of only three covered bridges still standing in that state. Trellis doors meet the weary traveler when crossing the charming, bucolic structure.

ABOVE: *New York is a city surrounded by water. Seventy-six bridges serve as valuable transportation links over these waterways. The oldest surviving bridge in New York City is the High Bridge. Built between 1837 and 1848, this bridge supported pipes in which water was carried from the upstate New York Croton Aqueduct to the 42nd Street Reservoir on Fifth Avenue (current site of the New York Public Library). The High Bridge has been declared a national landmark.*

ABOVE: *The Bridgeport Covered Bridge is the longest single-span covered bridge in the world. Built over California's South Fork Yuba River in 1862 for the Virginia City Turnpike Company, this bridge was an important link between the Comstock Lode in western Nevada and San Francisco. The outline of the interior wooden arch is clearly expressed on the exterior cedar shingles. The builder, David Wood, could never have anticipated the beautiful way in which the bridge has aged.* **OPPOSITE:** *The Wire Bridge spanning the Carrabesset River at New Portland, Maine, is still standing proudly. Crossing this wood plank–decked suspension bridge today, the traveler will have a nineteenth-century experience. The river flows swiftly below and the bridge sways gently in a stiff New England breeze. The towers are shingle-sided and the anchorages are made of large stones. During the period when this bridge was built, bridges were homegrown phenomena, and familiar and handy materials were always utilized.*

ABOVE: *When it was completed in 1813, the Casselman Bridge at Grantsville, Maryland, was the largest stone-arch bridge in the United States. This beautifully detailed and proportioned structure over the Casselman River is perfectly blended with its surroundings and makes the most of its eighty-foot (24.3m) span.*

OPPOSITE BOTTOM: *The Cornish-Windsor Covered Bridge is a double-span midpier structure that connects Vermont to New Hampshire over the Connecticut River. Built in 1866, after a century the bridge was in desperate need of strengthening due to a sag in the structure. The picture shows the success of "beefing up" the beams in place and restoring the original arching.* **ABOVE:** *John A. Roebling's oldest extant suspension bridge, the Delaware Aqueduct, was built in 1849 and has survived the ages; it is currently being renovated by the National Parks Service. Three midriver piers support the suspension cables, which are hidden by the high walkway of the aqueduct. The unusual structure that hides the cable was later clearly expressed and exposed in Roebling's technological wonder, the Brooklyn Bridge.*

LEFT: *The second-longest single-span covered bridge in the world, the Blenheim Bridge over the Schoharie Creek in North Blenheim, New York, is 210 feet (64m) long. Built in 1855 by Nicholas Montgomery Powers, the bridge is "double barreled" and uses a truss system patented in 1830 by Colonel Stephen H. Long.*

BELOW: *The Bollman Truss Bridge at Savage, Maryland, is an excellent example of a nineteenth-century iron bridge used extensively by the railroads to assist in their rapid expansion across North America. The white corner boxes are restored wooden housings protecting the eyebar and post connections. The bridge spans the Little Patuxent River and was built after the Civil War in 1869.*

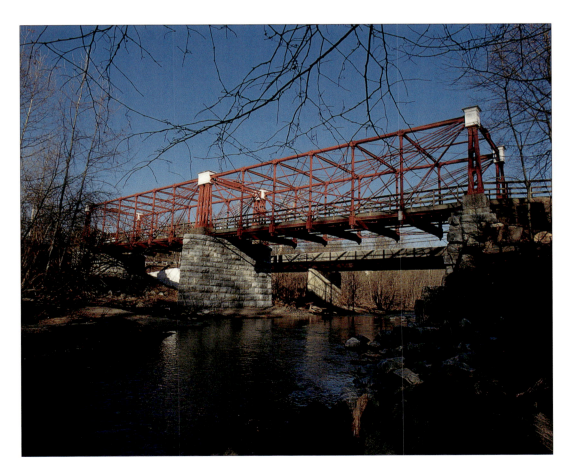

ABOVE: *This detail of Central Park's Bridge #28 shows the elegant art nouveau styling incorporated into many of the wrought-iron bridges designed by Calvert Vaux and Jacob Wrey Mould.* **RIGHT:** *New York City's Central Park is home to five beautiful and rare cast-iron park bridges. Designed by Calvert Vaux and Jacob Wrey Mould, this fine collection of decorative and graceful arched bridges is the oldest in America and is the perfect complement to Frederick Law Olmsted's landscape design. The Reservoir Bridge Southwest stretches seventy-two feet (21.9m) over the equestrian path.*

LEFT: *Elegant is the only word to describe Bow Bridge, the oldest wrought-iron girder bridge in the United States and the best New York City's Central Park has to offer. Dating from 1862, this cast-iron pedestrian bridge, with its gentle arch spanning the lake, provides a romantic setting with the cityscape beckoning beyond. The handsome balustrade with cinquefoils was designed by Calvert Vaux and Jacob Wrey Mould.* **ABOVE:** *The Pine Bank Arch was named for the grove of white pines growing on the nearby embankment. The first of five cast-iron pedestrian bridges designed by Calvert Vaux and Jacob Wrey Mould in the beautiful environs of New York City's Central Park, the Pine Bank Arch was built in 1861. This picture shows the beautiful restoration that was completed in 1984, in which the deck and missing elements of the balustrade and cinquefoils were replaced.*

ABOVE: *A beautiful wooden covered bridge in Amish country is the perfect backdrop for a horse-drawn carriage casting a shadow at sunset on this structure. Herr's Covered Bridge is located at Soudersburg, Pennsylvania.*

LEFT: *The Kapell Bridge in Lucerne, Switzerland, was built in 1333 over Lake Lucerne. Maintained in its original state, this covered wooden bridge has many paintings of medieval life and famous people hanging on interior panels. A large stone water tower guards one end of the structure.* BELOW: *Albert Goodwin's painting of this bridge— The Bridge of the Dance of Death—hangs at Christie's in London.*

OPPOSITE BOTTOM: *London Bridge is falling down? Not this bridge! This is the bridge that replaced the original London Bridge of nursery rhyme fame, built in 1209. John Rennie designed this bridge in 1824, and it was completed by his son, Sir John Rennie, seven years later. In 1968 the entire structure was moved block by block to Havasu City, Arizona, by developer Robert McCullock to anchor his desert community with a tourist attraction that has proven to be very popular indeed.*

LEFT: *The Haverhill Bath Bridge is the oldest covered bridge in the United States that is still in continuous use. This 278-foot (84.7m) double-span structure was completed in 1827. The bridge spans the Ammonoosuc River in New Hampshire, between the towns of Bath and Haverhill.*

ABOVE: *The Smith Rapids Bridge, in Chequamegon National Forest, Price County, Wisconsin, sports an unusual latticework system that allows plenty of light into the bridge's interior.*

RIGHT: *Built in 1697, the Frankford Avenue Bridge (seen here undergoing repairs) is the oldest bridge in continuous use in the United States. This bridge is a stone-arch structure that crosses Pennypack Creek in Philadelphia, Pennsylvania. Originally part of the King's Road, which was to connect Philadelphia to New York City, this three-span structure measures a scant seventy-five feet (22.8m).*

OPPOSITE BOTTOM: *The Choate Bridge is the second oldest stone bridge in the United States. Constructed in 1764 over the Ipswich River in Ipswich, Massachusetts, it was named for its builder, John Choate. The total length is over eighty feet (24.3m), with two thirty-foot (9.1m) elliptical stone arches supported by a midspan stone pier.* **ABOVE:** *The Barrickville Covered Bridge is a rare two-lane bridge built in 1853 over the Buffalo Creek in Barrickville, West Virginia. The necessary width for two-lane traffic is achieved through two unique features: the double roof and the double clerestory.*

SPANNING
THE ALPS

THE MAJESTIC AND SOARING ALPS SERVE AS

AN APPROPRIATE BACKDROP TO THE NEWEST

BRIDGE FORM OF THE TWENTIETH CENTURY:

PRESTRESSED- AND REINFORCED-CONCRETE

BRIDGES. REINFORCED CONCRETE IS THE

RESULT OF A BLISSFUL MARRIAGE OF TWO

ANCIENT MATERIALS: "ARTIFICIAL STONE"

AND IRON. ARTIFICIAL STONE WAS A MIXTURE

DERIVED FROM THE ROMANS, WHO COM-

BINED POZZOLANA, A NATURAL CEMENT,

WITH SAND AND LIME TO CREATE THE FIRST

NATURAL CONCRETE. THE PANTHEON IN

ROME, WITH ITS OPEN COLUMNLESS GALLERY,

IS A TRIBUTE TO THIS MATERIAL.

One early pioneer of reinforced concrete was François Hennebique, who designed the first reinforced-concrete arch bridge, the Pont de Chatallerault. This impressive bridge was built in 1898 over the Vienne River with a span of 172 feet (52.4m). This was the longest reinforced-concrete span constructed in the nineteenth century. The Pont de Chatallerault was eventually overshadowed by a bridge built over the Ourthe River at Liège, which opened in conjunction with the 1905 Paris Exhibition. Spanning 180 feet (54.8m), this reinforced-concrete arch with sculptural curves was a fine example of the possibilities that loomed in the future.

On February 6, 1872, a structural engineer with a vision as profound as Roebling or Eiffel was born. He was Robert Maillart, a man who revolutionized the use of concrete; almost all of the reinforced-concrete bridges that he created are still standing today. So beautiful are his bridges that the Museum of Modern Art in New York City gave Maillart an unprecedented one-man show in 1947. The rest of the art world was still not prepared, however, to accept an engineered structure as a work of art. Maillart single-

PAGES 60–61: The Val Tschiel Bridge, completed in 1925, is one of the elegant structures designed by Robert Maillart. This bridge spans the Val Tschiel ravine in the Alps. It has an arch-rib construction with a stiffening roadway, a thin arch shell, and transverse bearing walls. These elements are perfectly integrated in this mountainous setting, and the Val Tschiel is a uniquely successful and beautiful bridge.

handedly transformed that viewpoint and gave rise to the idea that utilitarian structures can indeed be considered and evaluated as art forms.

Maillart's first three-hinged reinforced-concrete arch, the Stanffacher Bridge, was built in 1899 over the Sibl River in Zurich, Switzerland. Ironically, the Stanffacher Bridge was covered in stone, which concealed the innovative three-hinged arch structure and retained the appearance of a traditional stone bridge. In 1901 Maillart eliminated the decorative stone facade for his Zuoz Bridge over the Inn River; he also utilized the structural walls to form the first-ever hollow-box reinforced-concrete bridge.

Always learning from his past creations, Maillart reached a new plateau in bridge design with his masterpiece over the Vorder Rhine at Tavanasa, Switzerland, in 1905. Maillart "sculpted" the concrete to visually express its three-hinged arch structure. This design was extremely radical and unconventional for the era; the public reception was chilly and Maillart was not to design another bridge of this form for more than twenty-five years.

The Tavanasa Bridge was destroyed by an avalanche in 1927. The replacement bridge, constructed nearby, was to be Maillart's most successful and dramatic bridge. With the Alps as a backdrop at Schiers, Switzerland, the Salginatobel Bridge spans more than 270 feet (82.2m) and is constructed entirely of concrete.

In 1913 Maillart developed a second bridge form: the deck-stiffened arch, which he used at Aarburg over the Aare River. Refining the design with each successive bridge, Maillart produced a masterpiece at Hinterfultigen, Switzerland, in 1933. For this bridge, the Schwandbach Bridge, Maillart eliminated the traditional stone abutments by using an extremely thin arch supporting a curving roadway; as a consequence, the continuous curving roadway is in perfect harmony with the structure. This bridge is a perfect

OPPOSITE: This picture shows two wonderful examples of a Devil's bridge. The old Devil's Bridge is in the foreground and the new Devil's Bridge is in the background. Both are located in the treacherous St. Gotthard Pass in Switzerland. The difficulty of constructing a bridge on the face of a mountain was once taken as a sign that the builder had made a pact with the Devil.

form and a work of art. Robert Maillart's bridge designs defied the complex bridge calculations of his contemporaries, and his work was generally neglected and underappreciated during his lifetime.

Eugène Freyssinet (1879–1962) was the second great man of modern concrete and the father of prestressed concrete. Freyssinet invented the idea of fitting a steel cable within a concrete beam and stretching the cable, thereby placing the concrete in high compression. Freyssinet also developed the process of prestressing concrete over a long period of time. The first bridge to have artificial forces applied to it was the Le Veurdre Bridge over the Allier River at Vichy, France.

Many remarkable improvements followed from Freyssinet's creation at Le Veurdre. In 1919 work was completed on a 315-foot (96m) double-ribbed arched span at Villeneuve-sur-Lot. In 1923 Freyssinet completed a 430-foot (131m) span over the Seine at Saint Pierre du Vauvray. When it was completed, this hollow arch was the longest concrete arch in the world. In 1930 Freyssinet completed an even longer reinforced-concrete arch at Brest, France; the arch extended an astonishing 592 feet (180.4m). The

Plougastel Bridge made Eugène Freyssinet famous throughout the world.

Freyssinet's first prestressed-concrete bridge was the Luzancy, built over France's Marne River in 1946. The prestressing of the concrete meant that the superstructure could be much thinner than had previously been required.

A notable designer who is following in the giant wake of both Robert Maillart and Eugene Freyssinet is Christian Menn. His Ganter Valley Bridge, built in 1980 near Brig, Switzerland, is a strikingly handsome concrete structure set in a valley among alpine mountain peaks. Two concrete pillars support a narrow, curving roadway with pre-

stressed cable stays encased in concrete. Encasement prevents corrosion, eliminates the problem of cable fatigue, and reduces stress. The Ganter Valley Bridge has altered the features of this valley as dramatically as John A. Roebling's Brooklyn Bridge altered the cityscape of New York City ninety-seven years earlier.

ABOVE: *The railroad bridges in the Glisons Mountains high in the Alps were constructed before the turn of the nineteenth century. The most dramatic and elegant is the Landwasser Viaduct near Filsur. Six proportioned arches, each standing more than two hundred feet (60.9m) high, lead the train track directly into a mountain tunnel. These beautiful stone structures are nearing their one hundredth anniversary.* **OPPOSITE:** *Built approximately two millennia ago, this ancient Roman double-arch bridge still stands proudly before the mighty Alps. Two thin elliptical arches span a stream in the Verzasco Valley of Switzerland. The arch was constructed without grout or mortar, and the stone voussoirs have held firm with only a minimum of maintenance.*

RIGHT: *The Salginatobel Bridge, located near Schiers, Switzerland, is a Robert Maillart masterpiece. Completed in 1930, this hollow-box three-hinged concrete arch was Maillart's longest span and his most famous bridge.* **BELOW:** *The scaffolding for the Salginatobel Bridge prefigures the elegance that the completed bridge will display.*

ABOVE: *Completed in 1913, Robert Maillart's second bridge form, the deck-stiffened arch with delicately light columns and stone abutments, spanned the Aare River in Switzerland. The arch carried the full load and, since it was not hinged, cracks appeared years later in the thin deck. This problem was resolved by the time Maillart built the Schwandbach Bridge in 1933.* LEFT: *Maillart's Aare River Bridge under construction.*

The Tavanasa Bridge was Robert Maillart's first bridge form and first masterpiece. Completed in 1905, this structure was the world's first reinforced-concrete three-hinged arch bridge that did not conceal its technology. Shocked by this expression of modern bridge building, the politicians and citizens of the Swiss Cantons did not permit Maillart to construct another three-hinged bridge for twenty-five years.

ABOVE: *The Toss River Bridge (1934) is one of Maillart's great deck-stiffened arch bridges. In the Toss River Bridge, he used a deep roadway curb to sufficiently stiffen the deck. In this bridge Maillart has eliminated the unnecessary heavy stone abutment that caused the cracking in the Aare Bridge in 1913. Maillart always learned from his mistakes, even twenty-one years later.*

RIGHT: *The Ganter Valley Bridge took Swiss bridge building to new heights. For this structure, Christian Menn designed a new bridge form that embeds prestressed cables in triangular concrete walls. The main span is more than 525 feet (160m) long and the support column is nearly five hundred feet (152.4m) high. The Ganter Valley Bridge is the longest span in Switzerland.*

LEFT: *The Schwandbach Bridge is an architect's and engineer's delight. The bridge required a minimum use of materials. It is pure of form, elegantly thin, and is in perfect harmony with its surroundings. Like all of Maillart's works, this 1933 masterpiece was ahead of its time. A moment of drama exists when its curving roadway is almost touched by a vertically rising arch at the center of its span. But the arch never actually touches the span—a stroke of genius.* **ABOVE:** *Robert Maillart's bridge over the Landquart in Klosters, Canton Grisons, Switzerland, built in 1930. Maillart utilized his second bridge form, the deck-stiffened arch, for this railroad bridge. In this structure, solid vertical cross walls integrate a stiffening beam deck with a thin arch.*

THE GREAT ARCHES AND CANTILEVERS

THE EXPLOSIVE GROWTH OF AMERICAN RAIL-ROADS DURING THE 1800S REQUIRED THE BUILDING OF THOUSANDS OF BRIDGES. IN THE LATTER PART OF THE CENTURY, AS THE CIVIL WAR WAS WINDING DOWN, THE CITY OF ST. LOUIS WANTED TO REESTABLISH ITS PREWAR DOMINANCE AS THE ECONOMIC POWERHOUSE OF THE MIDWEST. ACCORDING TO ITS CITIZENRY, WHAT ST. LOUIS NEEDED IN ORDER TO COMPETE EFFECTIVELY WITH CHICAGO, ITS NEIGHBOR TO THE NORTH, AS A TRANSPORTATION HUB AND COMMERCIAL CAPITAL, WAS A BRIDGE ACROSS THE MISSISSIPPI RIVER.

The people of St. Louis also wanted a symbol to proclaim their resurgence (a symbol they were not to find until one hundred years later in Alvar Alto Saarinen's Gateway Arch). To fulfill their goals, the city turned to a local businessman, James Buchanan Eads, who had designed and built ironclad ships for the Union Navy during the Civil War and who strongly believed in the need for this bridge from both a spiritual and a practical orientation.

Eads designed an innovative triple arch to be constructed from a newly available material: steel. This tough new material had just recently become commercially available through the development of the Bessemer Process and the combined contributions of Henry Bessemer, William Kelly, and Robert Mushet, all of whose work enabled steel to be made cheaply and in large quantities for the first time. Bessemer steel was made commercially for the first time in the United States in 1864 by William F. Durfee and his associates in Wyandotte, Michigan.

Steel is composed of iron alloyed with small percentages of carbon and often with other metals (such as nickel, chromium, and manganese), and has a hardness and a resistance to rust that opened up exciting new possibilities for bridge construction. No arch had ever been built to a length of more than four hundred feet (121.9m); the use of steel enabled Eads to make all three of his arches more than five hundred feet (152.4m) long. This structure was to be Eads' only bridge, and it was indeed a masterpiece. Completed in 1868, the Eads Bridge was the first bridge constructed of steel and was the longest arch (510 feet [155.4m]) in the world.

Two decades later, another new bridge form arose, this one on the Firth of Forth, near Edinburgh, Scotland. Benjamin Baker developed the cantilevered bridge truss, enabling spans to withstand heavy railroad loading. Baker's Forth Bridge had two steel spans of 1,710 feet (521.2m) each. In 1890 the opening ceremonies were attended by Gustave Eiffel and the Prince of Wales, among others. Like James Buchanan Eads, Benjamin Baker produced only one major bridge, but it too was a masterpiece. Both Eads and Baker introduced steel into new bridge design forms in the nineteenth century and made contributions that strongly influenced the course of twentieth-century bridge building.

The first true master bridge builder of the new century was Gustav Lindenthal, who served as bridge commissioner for New York City from 1902 to 1903. Lindenthal was a prolific designer who worked in a variety of design forms; he is responsible for the following spans: the Monongahela Bridge (1883), a steel truss bridge in Pittsburgh, Pennsylvania; the Williamsburg Bridge (1903), a suspension bridge in New York City; the Queensboro Bridge (1909), a cantilever bridge in New York City; the Sciotoville Bridge (1918), a continuous truss bridge; and the Hell Gate Bridge (1917), a steel arch construction in New York City.

Lindenthal's Hell Gate Bridge is a four-track railroad bridge for the New York, New Haven, and Hartford Railroad. The heavyset steel arch has a span of 1,038 feet (316.3m) and was completed in 1917. The power and visual weight of this structure is enhanced by two massive stone towers that are purely decorative. Critics either love or hate the Hell Gate Bridge, largely because of the addition of these nonfunctional stone towers. A good comparison span would be New Jersey's Bayonne Bridge built over the Kill van Kull (*kill* is a Dutch word meaning stream, channel, or creek) in 1931 by Othmar H. Ammann. This structure is also a steel arch, and its span is even longer: 1,675 feet (510.5m). However, it lacks the decorative stone abutments and towers of the Hell Gate Bridge. The Bayonne Bridge is "pure" structure, and no attempt has been made to hide or decorate its visual form.

In 1904 work began on the Quebec Bridge spanning the St. Lawrence River in Quebec, Canada. Due to many setbacks and delays, including a deadly structural failure that killed seventy-five workers, it took thirteen years to complete this structure. At 1,801 feet (548.9m), the Quebec Bridge, completed in 1918, is the longest clear-span cantilever bridge in the world.

The need for a permanent bridge in Calcutta, India, to replace a temporary pontoon bridge built in 1874 resulted in the selection of a cantilevered structure with a required span of fifteen hundred feet (457.2m). Until the new structure, named Howrah, was completed in 1943, the citizens of Calcutta continued to use the timber pontoon bridge built by Sir Bradford Leslie. Although the pontoon bridge had been built as a temporary solution and was expected to last only twenty-five years, it was used for sixty-nine years, at least partly because of the particular design problems presented by the site.

The Hooghly River was a tidal basin that rose more than twenty feet (6m) when the tide came in, which made the pontoon bridge rise and fall dramatically with the tides. Being a hinged bridge, the entry to the

PAGES 74-75: *The Sydney Harbour Bridge is a beautiful steel arch that dominates the harbor of this Australian port city. Similar to Gustav Lindenthal's Hell Gate Bridge, the Sydney Harbour Bridge has a pair of entry towers leading to a span of 1,650 feet (502.9m).* OPPOSITE: *James B. Eads built the first large-use steel bridge across the Mississippi River—the Eads Bridge—in St. Louis in 1874. Pneumatic caissons were sunk into the muddy Mississippi to secure a proper foundation for the bridge.*

pontoon, at these times, became so steep as to be nonnegotiable. The old pontoon bridge having finally outlived its usefulness, a contract for a new bridge was signed in 1936 with the Cleveland Bridge Company of Darlington; a cantilevered design was selected to help overcome some of the design complexities present at the site.

One of the unique advantages of a cantilevered bridge is that it does not require the sinking of any foundation piers into the middle of a raging river. Attempting to sink a pier in a river tidal basin such as the Hooghly would be an extraordinarily difficult task and one to be avoided at all costs. To construct a cantilevered bridge, two massive anchorage towers are constructed on each shore and a cantilevering structure is extend-ed from each tower. The two cantilevered arms are then joined by a span that is floated out and erected using jacks. Once in place, this suspended section of the span completes the structure.

The cantilevered bridge solution is not, however, without its own set of problems and complexities. It requires, first of all, perfect alignment of all structural members—a critical factor in the final construction of a cantilevered bridge. Moreover, since steel expands with heat and contracts with cold, final closure must be scheduled with the seasons. This further complicates the completion schedule, particularly in the Indian climate.

At Howrah, the completion of the final structure proved to be particularly dramatic. A half-inch (1.2cm) gap remained in the final connection of the structure, which was made up of 26,500 tons (24,062t) of steel and had a span of fifteen hundred feet (457.2m). That weight translated into fifty-three million pounds (24 million kg) and the dimension into eighteen thousand inches (45,720cm). At 9:00 A.M. on December 30, 1941, the drama and the temperature were both mounting. In a few hours, as the sun and the temperature continued to rise, the builders would no longer be able to join the half-inch (1.2cm) gap. The sixteen hydraulic jacks, each with a load capacity of eight hundred tons (726.4kg), were set up to move the separate parts of the structure and so close the final half-inch (1.2cm). The connection was successfully completed and the Howrah Bridge opened to vehicular traffic in February 1943.

ABOVE: *The Firth of Forth Railway Bridge near Edinburgh, Scotland, completed in 1890, was the masterpiece of Benjamin Baker. Two cantilevered spans of 1,710 feet (521.2m) each surpassed the Brooklyn Bridge as the longest span in the world. Baker pioneered the cantilevered bridge form that served the heavy-duty needs of modern railroads, and he was knighted for his efforts at the Forth.* **RIGHT:** *The Firth of Forth Railway Bridge under construction.*

ABOVE: *One of many New York City area bridges built by Othmar H. Ammann, the Bayonne Bridge in Bayonne, New Jersey—a majestic 1,675-foot (510.5m) parabolic arch span—was opened in 1932; it linked Newark Bay to New York's Upper Bay. High-strength manganese alloy steel was used for the first time for ribs and rivets. Ammann was a pioneer of "pure" structural design, and he eliminated the ever-present stone piers and towers that had defined the image of "the bridge" in the eyes of the general public.*

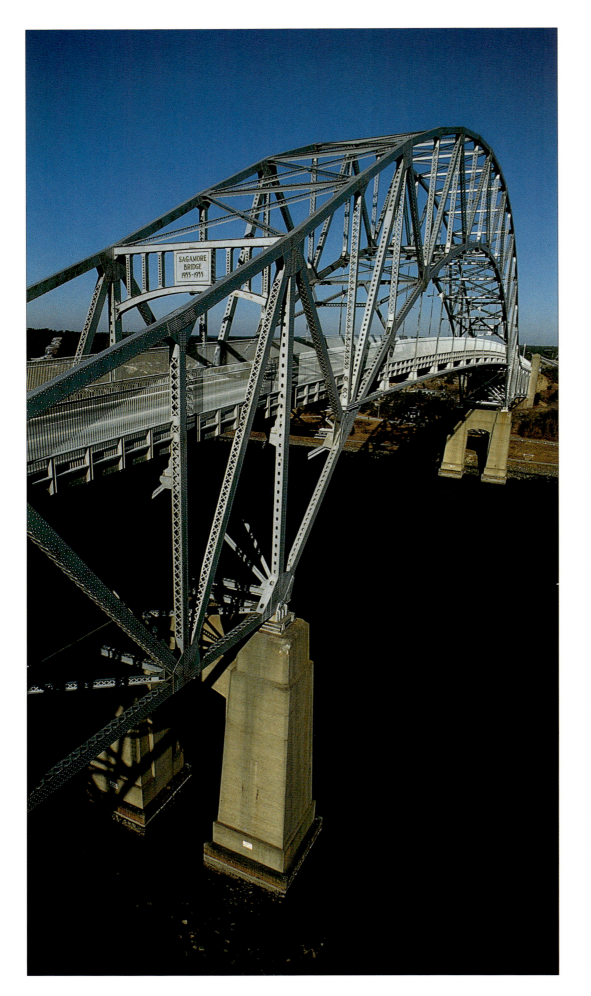

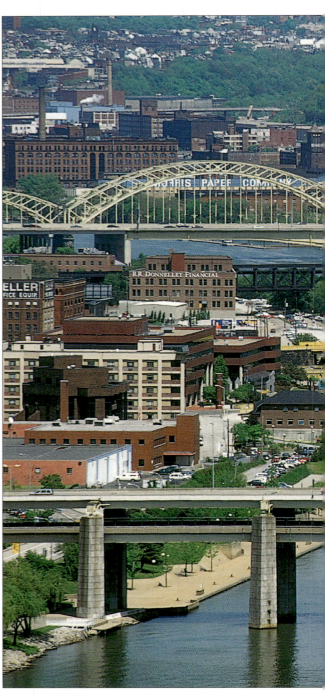

LEFT: *The Sagamore Bridge, built in 1935 outside Bourne, Massachusetts, is a single steel-arch structure that has a center span of about six hundred feet (182.8m). This bridge is near the end of Cape Cod and provides an important vehicular link to the popular vacation area each summer.*

BELOW: *The city of Pittsburgh, Pennsylvania, is situated at the confluence of three rivers, the Ohio, Allegheny, and Monongahela rivers. Geographic necessity created a need for many short structures with spans under one thousand feet (304.8m) long. A varied array of bridges cross the Allegheny River.*

Looking upstream, they include:

- •Ft. Duquesne Bridge
- •Sixth Street Bridge
- •Seventh Street Bridge
- •North Canal Street Railroad Bridge
- •Veterans Memorial Bridge—Interstate 579

ABOVE AND RIGHT: *The Queensboro Bridge, popularly known as the 59th Street Bridge, was completed in 1909. Gustav Lindenthal, chief engineer of New York City's Bridge Department, chose Blackwell's Island as the site for the stone piers of the longest continuous cantilever bridge to be built at that time in the United States. This location was John Roebling's alternate site for the Brooklyn Bridge. The unequal spans on either side of the island are 1,182 feet (360.2m) and 984 feet (300m), respectively. The total length of the bridge is 3,825 feet (1,165.8m).*

OPPOSITE TOP: *The Yaquina Bay Bridge is composed of several steel and reinforced-concrete arch spans that vary in length from 160 to 265 feet (48.7 to 80.7m). The main channel is six hundred feet (182.8) long and has a 226-foot (68.8m) clearing above the bay. The total length is more than thirty-two hundred feet (975.3m). Located at Newport, Oregon, the Yaquina Bay Bridge is an important link in the Oregon Coast Highway on U.S. Route 101. The chief engineer was Conde B. McCullough, and the bridge was completed in 1936.* **OPPOSITE BOTTOM:** *Both weighty and graceful, the arch of the Bridge of the Americas carries traffic across the Panama Canal near the canal's mouth on the Gulf of Panama. An essential link \in the Inter-American Highway, the Bridge of the Americas connects the northern and southern American continents.* **ABOVE:** *Draw Bridge, Seattle, Washington.*

Drawbridges are short-span low-clearance bridges that lift up on one edge or in the middle to allow river traffic to pass. Motorized gears, usually controlled from adjacent towers on a twenty-four-hour basis if traffic demands, are used to lift the parts of the bridge. The use of counterweights makes the motor's job easier and reduces the size of the motor required. However, the resulting structural appendage looks strange and awkward.

ABOVE: *Lift bridges have short spans that rise vertically up and down tracks on two towers situated on opposite piers. Designed to accommodate river traffic on busy waterways, lift bridges can have longer spans than drawbridges. This lift bridge in Chicago, Illinois, can rise more than 150 feet (45.7m).*

LEFT TOP: *Gustave Eiffel would have been pleased by the simplicity and elegance of the trestle piers that support the oil and gas pipeline running across the tundra in Lethbridge, Alberta. Leveling out the run through valleys and over mountains, stretches of this trestle are the world's longest and highest. This picture shows the setting Canadian sun shining through the slim steel struts as the structure heads toward the Arctic.* LEFT BOTTOM: *Over many millennia, the Snake River has carved out a deep gorge near Twin Falls, Idaho. Rising more than 486 feet (148.1m) above the Snake River, this Perrin member arch truss spans the river with minimum effort. Vertical columns extend from the arch truss to support the roadway above.*

OPPOSITE: *Opposite: The Jacques Cartier Bridge, spanning the St. Lawrence River in Montreal, has a total length of 11,236 feet (3,415.7m). Approximately four million rivets were used to assemble the imposing cantilever structure, which was built using 33,267 tons (30,179t) of steel. Construction began on May 26, 1925, and the bridge was officially opened on May 24, 1930.*

ABOVE: *The Dom Luis Bridge, which spans the Douro River at Oporto, Portugal, was constructed at the beginning of the twentieth century. The Dom Luis Bridge is a double-decker vehicular bridge and is very similar to its famous neighbor, the Pia Maria Bridge.*

ABOVE: *The Howrah in Calcutta, India, opened to traffic on February 28, 1943. The fourth-largest cantilever bridge in the world, the Howrah is composed of three spans, the longest of which is nearly fifteen hundred feet (457.2m) long. The structure also has two footpaths, one on each side of the roadway. A total of twenty-seven thousand tons (24,493t) of steel went into this immense structure.* **RIGHT:** *Gustav Lindenthal's masterpiece, the Hell Gate Bridge, was completed in 1917. The two-hinge spandrel-braced arch is a massive railroad bridge made up of eighty thousand tons (72,640t) of steel. Monumental stone abutments guard the approach on either side of the East River at a point in the river called Hell Gate.*

THE MIGHTY

AND

MAGNIFICENT

SPANS

THE STORY OF BRIDGES IS AS MUCH THE STORY OF OUTSTANDING INDIVIDUALS AS IT IS OF IMPROVEMENTS IN AVAILABLE MATERIALS AND STRUCTURAL DESIGN. BUILDING A BRIDGE REQUIRES A PERSON WITH A SPECIAL HEART, AN ALL-CONSUMING DRIVE AND DETERMINATION, AND, ABOVE ALL, A VISION.

During the Middle Ages, a journeyman bridge builder and visionary named Bénézet convinced the townspeople of Avignon, France, that he was on a divine mission to build a bridge for them across the Rhone River. The legendary Brotherhood of Bridge Builders (Frères Pontifes), headed by Bénézet, had constructed bridges throughout Europe with this same visionary zeal. The town officials of Avignon, however, were not convinced that a bridge could be built across their swift river. Nevertheless, Bénézet managed to convince the townspeople of the wisdom of his plan, and finally both the citizens and their officials rallied behind him with funds and support.

Bénézet's idea was to build the first masonry bridge in Europe since Roman times—the first such bridge built in Europe for almost one thousand years. His creation was technologically advanced for its day and quite elegant. The lightness of the elliptical arched spans probably led his critics to proclaim its imminent destruction. The obstacles Bénézet faced in the construction of his bridge at Avignon—to name but a few—included transporting heavy stone blocks

PAGES 92–93: Built in 1964, the Forth Road Bridge, located in Lothian, Scotland, is a suspension structure spanning 3,301 feet (1,006.1m). Its historic cantilevered namesake, the Forth Bridge over the Firth of Forth, is a railroad bridge built seventy-four years earlier. Suspension bridges such as this one are perfect for long spans and lighter vehicular traffic, while the sturdier cantilever structure can support the much heavier loads of the railroads.

from a distant quarry, choosing a suitable bridge site to best resist the rapid current, constructing several piers in the middle of the swiftly flowing river, raising sufficient construction funds, and convincing the many skeptics that his bridge would not fall.

In 1181 Bénézet died, just three years before completion of his bridge, and he was buried in the chapel that was constructed on the bridge. Today only the framework of four original arches remains of a bridge that survived the ravages of war and the unrelenting current of the Rhone River. Before its demise, the bridge had successfully served the citizens of Avignon for close to six centuries and had proven Benezet correct in his all-consuming vision.

Seven centuries later in the United States, engineer and bridge builder John Augustus Roebling, true to the nature of all great bridge builders, exhibited the same sort of determination, zeal, and vision. These were precisely the qualities required to build a daring structure across the East River connecting the great cities of New York and Brooklyn (an incorporated city in its own right from 1834 until 1898, when it was absorbed by New York City).

Roebling's dream was to build the longest suspension bridge in the world. He wanted to build a "Great Bridge" with a clear span across the East River without any intermediary piers in midstream. This would require a span of close to sixteen hundred feet (487.6m), a distance never before attempted. Roebling envisioned two bridge trains on a continuous cable carrying passengers over the bridge in less than five minutes. On the same deck, on either side of the trains, two vehicular traffic lanes would permit wagons and carriages to travel in each direction. Above the trains an "elevated promenade" would provide a way "for people to stroll across the bridge on fine days in order to enjoy the beautiful views and the pure air."

Roebling's vision included twin towers, each with a pair of Gothic arches one hun-

dred feet (30.4m) high. His proudest achievement would be to suspend this magnificent and daring structure with steel wire cables and to tie everything together with steel inclined stays and steel vertical suspenders, thus providing a stability and strength that no critic would be able to deny. This would be the first bridge utilizing steel cables in its construction.

Like Bénézet centuries before him, Roebling faced obstacles as grand as the project he intended to build. The two massive stone towers, each weighing over sixty-seven thousand tons (60,836t), had to have firm and solid foundations. These colossal towers, each rising to a height of 276 feet (84.1m), would be taller than any existing buildings in the area and thus would dominate the skylines of both New York and Brooklyn. Roebling solved his problems and pushed forward with plans for construction, but did not live to see the project through to completion.

Before this visionary engineer had the opportunity to see his towers rise above the skyline, he died as the result of a freak accident while surveying for the towers. Trying to avoid the impact of an incoming Fulton Street Ferry on the East River pier on which he was standing, Roebling caught his foot between a rope and the timbers of the pier, resulting in the amputation of several toes. While the amputation didn't kill him, his attempt at home remedies did. Thinking he could cure himself, he continually ran cold water over his wound, despite the objections of the many doctors whom he continued to dismiss and replace. His wound festered and tetanus set in, resulting in lockjaw. Roebling suffered immeasurably over an extended period of time, experiencing terrible pain; his strength of will astounded those physicians still on call. Like Bénézet at Avignon and Peter of Colechurch at the London Bridge, Roebling died before the completion of his masterpiece.

Roebling's thirty-two-year-old son, Colonel Washington A. Roebling, who had been assisting his father with the Brooklyn Bridge project, was promoted to chief engineer and undertook the completion of the work. The new chief engineer devised a pressurized timber-, tin-, and iron-clad caisson that would support the twin towers. As the excavations continued, the pneumatic caissons, which were the largest in the world, were slowly sunk into the soft riverbed. Washington Roebling described the caissons as "diving bells" with open bottoms that formed an airtight pressurized work chamber. Workers descended into the caissons through airlocks in order to dig out the soft riverbed, thus enabling the continued lowering of the caissons by the massive stone weight being placed upon them. In this way, the caissons eventually reached a solid rock foundation where they could be anchored and filled with concrete.

One obstacle that Roebling faced in realizing his father's dream was the debilitating condition then called "caisson disease" (now known as the bends), an affliction unknown at that time, as no other construction project had required divers to go so deep into a riverbed. At the same time that this project was proceeding, workers at James Buchanan Eads' St. Louis bridge project were experiencing similar symptoms as a result of deepwater caisson work. Eads and Roebling corresponded about this problem, and Roebling eventually inspected the caissons in St. Louis; but no resolution was forthcoming since neither builder understood the underlying causes of the bends. This debilitating condition, it was later learned, resulted from too-rapid decompression experienced after exiting from the high pressure of the caissons. In this situation, nitrogen bubbles would get caught in various parts of the body and cause symptoms varying from heart failure to crippling paralysis. Moreover, caisson disease mysteriously affected some workers, while leaving others unscathed; those who worked in the caissons regularly seemed less likely to be affected

than those who ventured into the caissons only occasionally.

Eventually, Washington Roebling was himself to succumb to a crippling and disabling paralysis as a direct result of caisson disease. After this condition took hold, Roebling was forced to watch the ongoing construction through field glasses from his Brooklyn Heights window. Under his direction, his wife, Emily Warren Roebling (sister of General G. K. Warren) and his assistants (Colonel Paine, C.C. Martin, and Wilhelm Hildendrand) completed the bridge in fourteen years.

The Brooklyn Bridge was formally dedicated and opened by President Chester A. Arthur on May 24, 1883. This bridge was one of the greatest engineering feats of the nineteenth century and one of the most inspiring. The successful completion of John Roebling's "Great Bridge" marked the beginning of the suspension bridge era.

John Roebling had predicted that the clear span of suspension bridges could be more than twice that of the Brooklyn Bridge. His prophecy became a reality in 1931, when the George Washington Bridge, built by Othmar H. Ammann, spanned the Hudson River with a clear span of thirty-five hundred feet (1,066.8m). This bridge was originally designed with stone-covered steel towers, but in the end the steel towers were left unsheathed, resulting in a "dense web" trussed tower that proved to be as elegant and as appropriate as the span itself. (The towers were left bare because the Great Depression had placed budgetary constraints on the project.)

At night, the reflection of the cable lights dancing on the Hudson River, shimmering like a string of pearls gently descending from the tips of the towers, is breathtaking. The natural setting also contributes to the elegance of the scene. To anyone sailing or driving up the Hudson River at dusk, the magnificent panorama of the stately George Washington Bridge presents itself between

the high bluffs of the Palisades on the New Jersey side and Washington Heights on the New York shore.

The original single deck of the George Washington Bridge is suspended from four thirty-six-inch (91.4cm) cables, each composed of more than twenty-six thousand wires that together are more than one hundred thousand miles (160,900km) in length. The cables were spun in place by a traveling mechanized spinner followed by a "cable squeezer." Wrapping machines completed the modern cable installation. In 1962, due to increased vehicular traffic, a second, lower deck was added to the George Washington Bridge. Fortuitously, Othmar Ammann's original design had planned for that additional deck.

Ammann was a master of the suspension bridge; his other bridges of this type were all built in the New York City metropolitan area: the Triborough Bridge (1,380 feet [420.6m], 1936); the Bronx-Whitestone Bridge (twenty-three hundred feet [701m], 1939); and the Verrazano Narrows Bridge (4,260 feet [1,298.4m], 1964), the longest suspension bridge in North America.

On the opposite coast of the United States, as early as the start of construction on the Brooklyn Bridge in the 1860s, discussions were underway about the construction of a bridge to span the Golden Gate of San Francisco Bay. The dream was not fulfilled until 1937, however, when a slender suspension bridge of forty-two hundred feet (1,280.1m) was completed. Like the Brooklyn Bridge, the original plans for the Golden Gate Bridge called for a foundation caisson to be sunk one hundred feet (30.4m) below water level in order to find a solid rock footing. Unfortunately, the eighteen-thousand-square-foot (1,674m²), eight-thousand-ton (7,264t) caisson was caught in a Pacific swell and wrecked before it could be placed in position. Consequently, the engineers working on the bridge switched to a cofferdam-designed foundation and poured sixty-five feet (19.8m) of concrete into it to complete the foundation.

The twin towers of the Golden Gate Bridge sail more than seven hundred feet (213.3m) into the air and have a very shallow deck-stiffening truss of only twenty-five feet (7.6m). The ratio of the bridge width of sixty-six feet (20.1m) to its span of forty-two hundred feet (1,280.1m) results in a very modern and slender width-to-span ratio of 1:63. In an extremely heavy gale-force wind of 120 miles per hour (193.1kph), the bridge was calculated to sway more than twenty feet (6m). This sway was deemed "acceptable" by the engineers in charge.

On November 7, 1940, another very slender suspension bridge, the Tacoma Narrows Bridge, made headlines when it was set in motion by a steady forty-mile-per-hour (64kph) wind. Heaving and twisting uncontrollably, the bridge finally collapsed when the hangers and cables snapped. The bridge, which was nicknamed "Galloping Gertie" after it was destroyed, should never have been built without the proper stiffening demanded by Roebling for all of his own suspension bridge designs; the engineers in charge of Gertie's construction should have been aware of this from Roebling's extensive writings on the subject.

Charles Ellet, Jr.'s Wheeling Suspension Bridge over the Ohio River, built in 1849, collapsed in the same fashion as "Galloping Gertie" only five years after its completion. This collapse should also have served as a warning to the designers and engineers of the Tacoma Narrows Bridge that proper stiffening was necessary. But for whatever reason, they chose to ignore this important safeguard measure. Both collapsed bridges were aerodynamically unstable and clearly required additional stiffeners.

Although their bridge was not as slender as "Galloping Gertie," the Golden Gate Bridge engineers installed recording devices to check on deck movements. In 1951 a steady storm with winds in excess of sixty miles per hour (96kph) produced a twelve-foot (3.6m) lateral sway and a ten-foot (3m) deck undulation. Wisely, the engineers installed additional lateral deck stiffeners below the roadway along the total length of the structure, thus eliminating any "acceptable sway." An alternative method of stiffening the Golden Gate Bridge would have been to install diagonal "inclined" stays, the structural signature of John A. Roebling.

Today, several new suspension bridges are under construction. The Tsing Ma Bridge in Hong Kong, scheduled for completion in 1997, will have a span of 4,518 feet (1,377m). Other suspension bridges under construction at the time of this writing include structures of record-breaking lengths. In 1996 the Store Baeit (East Bridge) in Denmark will be completed; it will have the world's longest span: 5,328 feet (1,623.9m). This record will be broken less than two years later, when the Höga Kusten (High Coast) Suspension Bridge in Vasternorrland, Sweden (modeled after the Golden Gate Bridge) supplants it as the longest. When completed in 1997, the clear span of this bridge will be close to six thousand feet (1,828.8m), and the twin tapered towers will rise six hundred feet (182.8m) above the water's surface. This record, however, will also be very short-lived. By 1998 the new Akashi Kaikyo Bridge in Akashi, Japan is expected to be completed with a span of 6,529 feet (1,990m).

Thus John A. Roebling's dream of a mile-long suspension bridge is close at hand—and soon to be surpassed. His critics, doubters, and detractors are now just interesting historical footnotes.

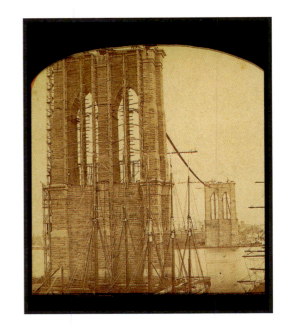

OPPOSITE: *John A. Roebling's nineteenth-century masterpiece, the Brooklyn Bridge dominates the entrance to the East River in New York City. This bridge epitomizes the suspension bridge. Massive granite stone towers rise more than 276 feet (84.1m) above the river. More than five thousand steel wires make up each of the four 15¾-inch (40cm) -diameter cables that support the bridge. The Brooklyn Bridge was designed by John A. Roebling but it was constructed under the supervision of his son, Colonel Washington A. Roebling, after the senior Roebling died from injuries sustained in an accident while surveying for the New York tower.* ABOVE: *The Brooklyn Bridge under construction.*

ABOVE: *Constructed during the Civil War and opened in 1866, the Cincinnati Suspension Bridge connecting Cincinnati to Covington, Kentucky, was the world's longest span—1,057 feet (322.1m) when it was completed. John A. Roebling designed the structure and his son, Washington A. Roebling, supervised the construction.*

With this bridge, the Roebling father-son team developed the necessary engineering techniques to construct suspension bridges. Stiffening trusses were added in 1895 and additional cabling was added in 1899.

OPPOSITE BOTTOM: *Spanning the Hudson River near West Point, New York, the Bear Mountain Bridge was finished in 1924.*

The 1,632-foot (497.4m) span was the longest suspension bridge when it was completed. Within seven years, it was surpassed by a rival bridge with more than twice its span (thirty-five hundred feet [1,066.8m]) when Othmar Ammann's George Washington Bridge opened to traffic a short distance downriver.

ABOVE: *The Throggs Neck Bridge is the only bridge to span Long Island Sound between the boroughs of Queens and the Bronx. Completed in 1961, this bridge spans eighteen hundred feet (548.6m) and is the most recent structure across the East River. The first bridge spanning this river was the Brooklyn Bridge, built in 1883.*

ABOVE: *As Bridge Commissioner of New York City, Gustav Lindenthal originally designed the Manhattan Bridge as an eyebar chain suspension bridge. He also commissioned the architects Carrère and Hastings to collaborate on the design. The design was changed to a wire cable suspension bridge when Lindenthal left his position before the bridge was built in 1909.* **RIGHT:** *This is the entrance to the Manhattan Bridge as it appeared in 1917.*

LEFT: *When the Manhattan Bridge first opened, the southern roadway was dedicated to trolleys, with only the northern roadway open to motor traffic.*

OPPOSITE: *The George Washington Bridge spans the Hudson River between the Palisades in New Jersey and Washington Heights in New York. Built in 1931, the George Washington Bridge is thirty-five hundred feet (1,066.8) long, more than double the previous record for suspension spans. It is Othmar H. Ammann's greatest achievement.* **RIGHT TOP:** *The longest suspension bridge in the United States, the Verrazano Narrows Bridge was designed by the firm of Ammann and Whitney and was opened to traffic in 1964. The total length of the bridge is seventy-two hundred feet (2,194.5m), with a clear span of 4,260 feet (1,298.4m). Dominating the entrance to New York Harbor, this bridge stands 216 feet (65.8m) above the water, thus providing a grand and majestic portal for ships entering and leaving the harbor.*

ABOVE: *The San Francisco–Oakland Bay Bridge was built in 1936. The two transbay bridge spans are 2,310 feet (704m) each. The total length of the West Bay crossing is ninety-two hundred feet (2,804.1m). The East Bay crossing has a total length of eleven thousand feet (3,352.8m). These are the only double-span suspension bridges in the world. This photograph provides a view of the West Bay span connecting San Francisco in the distance to Yerba Buena Island in the middle of the bay.*

ABOVE: *One of the most famous bridges in the world, the Golden Gate Bridge has a magical Hollywood name enhanced by a coat of red oxide paint that seems to glow during sunsets. The 746-foot (227.3m) art deco towers sail high above San Francisco Bay. Built in 1937, the 4,260-foot (1,298.4m) span was the longest in America until 1964.* RIGHT: *The Golden Gate Bridge under construction.*

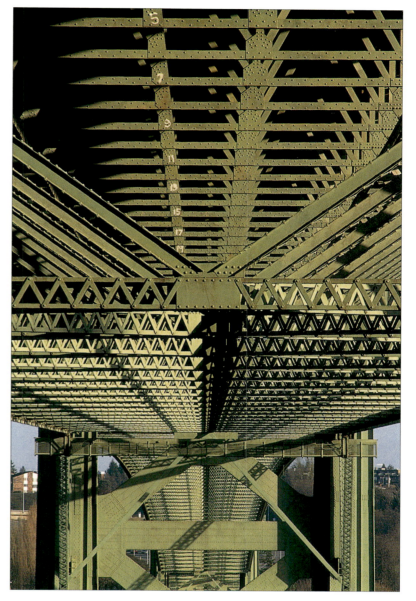

ABOVE LEFT: *The original Tacoma Narrows Bridge, known as "Galloping Gertie," collapsed due to its inflexibility in the wind. The replacement bridge, erected in 1950, rectified that problem with a deep space truss system that allows the wind to pass through the structure.* **ABOVE RIGHT:** *This picture clearly shows the strengthening of the underside of the bridge with an extensive array of struts and trusses*

LEFT: *When Charles Ellet, Jr.'s innovative suspension bridge was constructed in 1849, spanning the Ohio River at Wheeling, West Virginia, it was the longest in the world, with a clear span of 1,010 feet (307.8m). A one-thousand-foot (304.8m) span was unheard-of in those days, and building one required both breakthrough technology and a personal determination to achieve. In 1854 the bridge collapsed in high winds, but was rebuilt.*

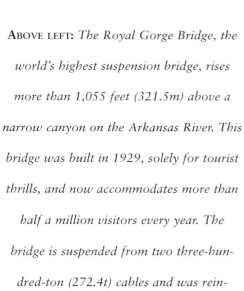

ABOVE LEFT: *The Royal Gorge Bridge, the world's highest suspension bridge, rises more than 1,055 feet (321.5m) above a narrow canyon on the Arkansas River. This bridge was built in 1929, solely for tourist thrills, and now accommodates more than half a million visitors every year. The bridge is suspended from two three-hundred-ton (272.4t) cables and was reinforced in 1983. A tram, its tower barely* visible in the photograph, at the center right below the bridge tower, opened in 1969. ABOVE RIGHT: *The painting* Spectators Packing the Hammersmith Bridge on Boat Race Day, *by Walter Graves, hangs in the Tate Gallery in London. Spectators crowd the suspension cables as if they were in the upper balcony in a theater, and all of them have a bird's-eye view of the boat races below.*

OPPOSITE: *Built in 1957, the Mackinac Straits Bridge crosses the straits between Michigan and its upper peninsula. The thirty-eight-hundred-foot (1,158.2m) bridge stands at the northern tip of mainland Michigan at the point separating Lake Huron and Lake Michigan. The slender deck is stiffened with a deep truss built below the superstructure.*

RIGHT: *Currently the longest suspension bridge in the world, the Humber Bridge spans 4,626 feet (1,410m). Built in England in 1981, the Humber will retain its "world's longest" title only until the completion of Store Baeit (East Bridge) in Denmark, which is scheduled for completion late in 1996. The Store Baeit will be 5,328 feet (1,624m) long, but will hold its record only briefly. By 1998 the Akashi Kaikyo Bridge in Japan, with a new record length of 6,529 feet (1,990m), will be completed.*

LEFT: *The Bosporus Bridge in Istanbul, Turkey, was completed in 1973. A second bridge over the Bosporus was opened in 1988. The original suspension bridge pictured here spans 3,524 feet (1,074.1m), while the newer Bosporus Bridge spans a length of 3,576 feet (1,089.9m).* **ABOVE:** *The Ponte 25 de Abril spans the Tagus River in Lisbon, Portugal. Glowing majestically at sunset, this suspension bridge spans 3,323 feet (1,012.8m). Opened to traffic in 1966, it is one of the longest bridges in Europe.*

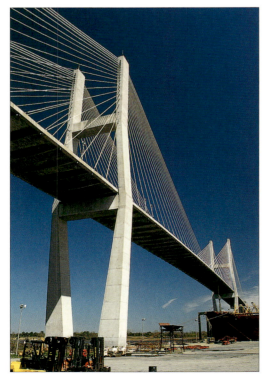

LEFT: *The Talmadge Memorial Bridge is a newly completed cable-stayed structure in Savannah, Georgia. The 1,100-foot (335.2m) clear span opened to vehicular traffic in 1991.* ABOVE: *Currently under construction in Hong Kong Harbor, the Tsing Ma Bridge is scheduled to open in 1997. The span of the bridge between the towers will be 4,518 feet (1,377m). This picture shows the progress to date, with the towers and suspension cables completed and the vertical stays and deck superstructure more than halfway done. By the time this book is published, vehicular traffic should be moving across the Tsing Ma.* OPPOSITE: *The Queen Elizabeth Bridge over the Thames River in Dartford, England, is a cable-stayed bridge. The cable-stayed bridge is a recent design development that uses only diagonally strung wire cables through steel towers. The simplicity and elegance of the design and structure astound the viewer.*

LEFT: *The Broward Bridge in Jacksonville, Florida, is the longest cable-stayed span in the United States. Built in 1988 at Dames Point, the modern wire cable span is thirteen hundred feet (396.2m) long.* ABOVE: *The dramatic and daring Cohlbrand Bridge in Hamburg, Germany, is a cable-stayed structure with a reinforced-concrete curvilinear roadway. Supported by diagonal stays splaying off a single maypole tower, the roadway "floats" between the tower supports in a most dramatic fashion.*

LEFT TOP: *The cable-stayed bridge over the Mississippi River at Luling, Louisiana, was opened in 1983. The 1,222-foot (372.4m) structure is an important link in Interstate Highway 310. The cable-stayed bridge minimizes structural elements and has simplified the construction of wire cable bridges.* LEFT BOTTOM: *The Great Bridge in Japan, called Minami Bisan-Sato, is the second-longest suspension bridge in the world. Completed in 1988, the 3,609-foot (1,100m) bridge has a truss space frame to stiffen the roadway deck.* OPPOSITE: *The Höga Kusten Bridge is a four-thousand-foot (1,219.2m) suspension bridge currently under construction in Vasternorrland, Sweden. The pylons are more than six hundred feet (182.8m) tall. This bridge, which will be completed in the summer of 1997, will serve as an important gateway to the Höga Kusten (High Coast) region. One up-to-the-minute feature of this bridge's construction is that a camera covers the project twenty-four hours a day and provides continually updated photographs to a website at www.connection.se/hogakusten/index.html. Visitors to the website are invited to use the camera's zoom feature for a closer view of the structure.*

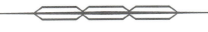

GLOSSARY

Bascule: A type of drawbridge counterweighted so that it can be raised and lowered easily. The term comes from a device that works like a seesaw; when one end is lowered, the other end is raised.

Cable-stayed bridge: A bridge constructed along a structural system comprised of a deck and continuous girders that are supported by stays. Inclined cables pass over or are attached to towers located as the main piers.

Caisson: A watertight enclosure used in underwater construction. The term is also used for a hollow box which is used as a floodgate at a dock or basin.

Cantilever: A projecting beam supported at only one end. The supported end is most often anchored to a pier or wall. The term is also used in construction to connote a large bracket projecting from a wall to support a balcony or cornice.

Cantilever bridge: A bridge whose span is formed by two beams projecting toward each other, sometimes with an extra section between them. Each beam is supported at only one end.

Cofferdam: A watertight temporary structure used in bridge building to keep water away from an area that has been pumped dry. It is used to create a dry section of a lake or river bed, allowing construction of bridge foundations unimpeded by water.

Eyebar: A metal bar, generally rectangular in cross-section, enlarged at each end for holes, or "eyes."

Pozzolana: Any substance added to cement to lend particular properties; substances may be natural or artificial.

Prestressed concrete: Concrete which contains steel cables, wires, etc. under tension. Used to lend greater strength to a structure.

Span: The distance between ends or supports. Also used to connote the part of a structure between two supports and, in bridge building, as a synonym for bridge.

Spandrels: The triangular space between the exterior curve of an arch and a rectangular frame or molding enclosing it. Also used to connote any of the spaces between a series of arches and a straight cornice running above them.

Truss: A rigid framework of beams, girders, struts, bars, etc. used to support a bridge. Truss bridges are comprised of lattices formed by straight members in a triangular pattern. The most common form of bridge truss is the Warren truss. Others include the Subdivided Warren truss, the Pratt truss, the curved-chord Pratt truss, the How truss, and the K-type truss.

Voussoirs: Wedge-shaped stones used to construct an arch or vault.

INDEX

BIBILIOGRAPHY

Billington, David P. *The Tower and the Bridge: The New Art of Structural Engineering.* Princeton, NJ: Princeton University Press, 1963.

"Bridges." *The 1996 World Almanac and Book of Facts.* Mahwah, NJ: World Almanac Books/Funk & Wagnalls, 1995.

DeLony, Eric. *Landmark American Bridges.* New York: American Society of Civil Engineers, 1992.

Ellis, Edward Robb. *The Epic of New York City: A Narrative History.* New York: Old Town Books, 1966.

McCullough, David. *The Great Bridge: The Epic Story of the Building of the Brooklyn Bridge.* New York: Touchstone/Simon & Schuster, 1972.

Naruse, Y. and T. Kijima, eds. *Bridges of the World.* Tokyo: Morikita Publishing Co., 1967.

Smith, Shirley H. *The World's Great Bridges.* Rev. ed. New York: Harper & Row, 1965.

Troisky, M.S. *Planning and Design of Bridges.* New York: John Wiley & Sons, 1994.

Wells, Rosalie. *Covered Bridges in America.* New York: William Edwin Rudge, 1931.

PHOTOGRAPHY CREDITS